POPULAR GUIDES TO GREAT PHOTOGRAPHY:

"where your photographic rewards begin"

Vol. 6
PHOTOGRAPHING WATER
3rd Edition

Fully Revised and Expanded
with around 200 illustrations

By

JOHN C. DOORNKAMP

Award Winning Photographer

Text and Photographs
© Copyright John C Doornkamp (2022)
All Rights Reserved

No part of this publication may be reproduced, stored in a retrieval system or transmitted in any form or by any means without the prior written permission of the author.

This book is for those who want to learn more and learn quickly about photographing water without "drowning in the deep end".

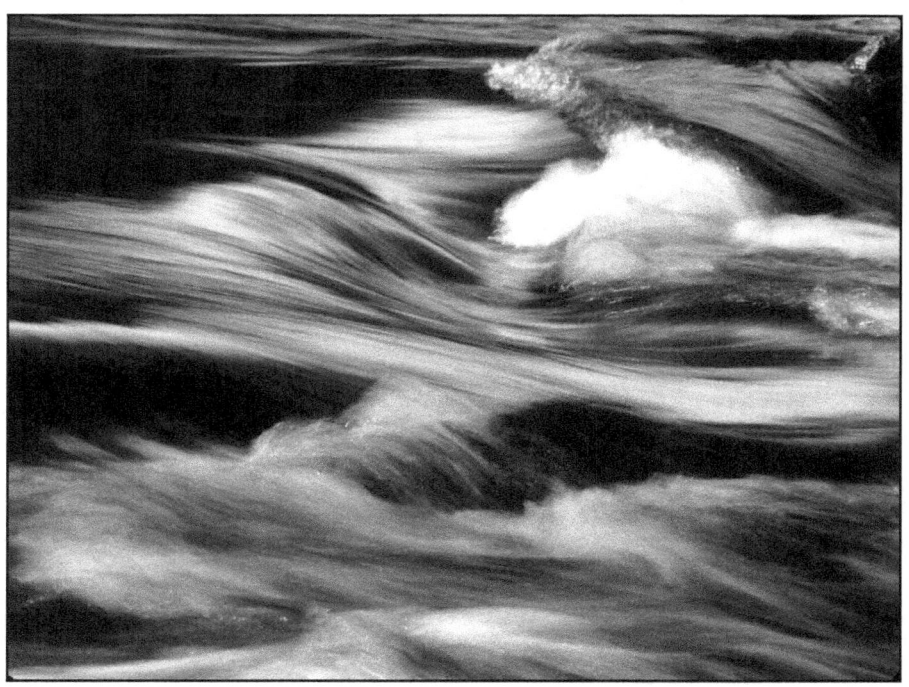

Table of Contents

PREFACE
1 INTRODUCTION
2 THE NATURE OF WATER
    Composition
    Shutter Speed
    Aperture
    Depth of focus
    Filters
    Viewpoint
    The Hydrological Cycle
    Exposure readings and water
3. CLOUDS
4. DEW AND RAIN
5. FLOWING WATER
6. FALLING WATER
7. BREAKING WAVES
8. STILL AND CALM
9. SALINE LAKES
10. STEAM
11. FROST, ICE AND SNOW
12. WATER IN PICTORIAL IMAGES
13. WATER AT WORK IN NATURE
14. WATER AS A LIVING ENVIRONMENT
15. HUMAN USES OF WATER
16. TECHNICAL NOTES
    - extending exposure time
    - HDR photography
WATER - A FINAL COMMENT.
QUICK SUMMARY
ABOUT THE AUTHOR

PREFACE

Waterfall, Milford Sound, South Island, New Zealand
(Exposure:1/30th sec at f/9, lens focal length 90mm, ISO 400)

Why have I bothered to write this book?

The first and second editions were well received.
Photographing Water has been one of my top selling titles. So, why put in all of the extra time and effort required to write a third edition? The answer is simple.

Despite the compliments and good reviews for the first two editions my own instinct was that it didn't really do justice to my passion for photographing water. The problem with having a passion is that you want to share it with others. I wanted to share more fully the richness of images that can be created from photographing water. Hence the inclusion of around 200 examples, many of which were not in the first edition.

I believe that we are better photographers if we understand our subjects. So, in this third edition there is a much fuller explanation of how water carries out its work in fashioning landscape. I have tried to include images that illustrate the nature of water as a process whilst still retaining a pictorial approach.

In addition, this third edition contains an expanded section on the subject of how people interact with water. How they treat it both as a resource and as a natural peril. Water can be used as a means of survival or as a decorative feature. Both lend themselves to photography. Reservoirs can be treated as lakes (for photographic purposes) and a fountain can be treated as falling water.

Water can also pose a threat to people and their property. The reaction is to build defences, and here again (as this book will show) these can provide a subject for our photography. Water is seen by many as a playground, and by others as something to be included in landscape design. Within all of this there are many opportunities for finding photogenic situations that will reward the skilled photographer. What is required is both the 'eye' for a photogenic subject, and the technical skill to record it well. This book is designed to help you with both.

Photographing water has for me become totally absorbing. Every success has brought its own thrill. None more so than the image at the head of this Preface. More on that in *Section 6 'Falling Water'*. My hope is that through the examples you will see in this book you will be stimulated to find your own satisfying water-related images. This does not come without some effort. Opportunities don't always fall into your lap. You have to take the trouble to go and look for them and recognise their picture-making potential. In order to realise that potential you will also require a good technique.

In this book I will tell you what has worked for me. There may be some ideas and methods that are new to you. Feel free to use them, and then improve upon them and make these your own. So, with all of this in mind, dive in and see for yourself. However, be warned, photographing water can become a passion!

1
INTRODUCTION

Photographers at Yellowstone National Park.

Ladybower Reservoir (Peak District National Park, England)
(Note: Taken with a wide-angle zoom lens set at a focal length of 12mm Exposure: 1/100th sec at f/11 using ISO400, tripod)

Whatever form it may take (and there are many), water is an unfailing attraction to photographers.

Photographing water can be a joy, and totally absorbing. It does not matter whether it takes the form of a spectacular waterfall, or just a quiet lake with a soft reflection – people will raise their camera and take a photograph. This spontaneous response to water means that we are not always as thoughtful as we might be about our photographic techniques.

This book is about getting those techniques right every time. This book is also about the way in which our techniques will differ according to the state in which water occurs.

Water can occur as a solid (e.g. glaciers and ice) or as a liquid. We see it in seas, rivers and lakes. Sometimes it is breaking in waves, falling, flowing, or lying still. Whatever its state, water is always an attraction.

We, as human beings interact with water, we need it to survive, we play in it, we use it for transport, and we can reflect upon its power and beauty. Water carries out work as an agent of erosion, transportation, and the deposition of sediment. Water is also a medium within which life abounds - whether as fish, mammals or some other life form. These, and more, are the subjects of water photography.

These different states are reflected in the Section titles. You can dip in and out of each Section or you can read the book from beginning to end. However you choose to approach this volume I know that there is something in it that is of value to you.

This book leads by example. The 200 or so photographs are designed to provide you with ideas. You may want to imitate my methods or you may want to use these as a starting point for your own approach to photographing water.

**Ladybower Reservoir and Glossop Road Bridge
(Peak District National Park, England)**
(Exposure: 1/15th sec at f/14 set at ISO400, tripod)

The previous two images illustrate two of the questions that arise when photographing water:
　'Do I take a wide shot or do I get in closer?'
and 'Do I include any foreground or do I just photograph the water?'

Ladybower is the lowest in a flight of four reservoirs where water is stored for the people of Sheffield, Nottingham and Leicester (in the U.K.). This is an example of water as a resource.

We all know that water can also be a terrifying hazard - when it floods, or is caught up in a tsunami, or simply freezes over and creates chaos in the process. Water, therefore, not only occurs in a variety of physical forms it also affects our lives in a variety of ways. As a result there is ample opportunity to create images with 'water' as the main subject.

Shooting the rapids…through choice.
(200mm telephoto, 1/100th sec, f/5.6, ISO100)
(It was important to wait for the subject to be facing the camera.)

Washing the car …. because he has been told to do so!
(Exposure: f/16 for depth of field, wide-angle at 35mm so as to include both the boy and the car).

Sometimes an image can become 'just another photograph of water' (yawn). When it starts to feel like that it may help to include a person interacting with that water. In this way you give the image an extra meaning and interest.

In the case of the kayak shooting the rapids not only does the turbulence of the water give a sense of power and force, it has more meaning because we can see that there is someone out there doing battle with the raging water.

Washing a car, on the other hand, is a mundane task. Here, however, the image becomes worthwhile because of the interplay between the spray of water and the sunlight reflected by the water droplets and the role of a person.

In these two examples we have two completely different photographic opportunities, as well as two different types of interaction between humans and water. The first (a river near the USA-Canada border) gives us an insight into the way that people can use a river. The nature of their activity is related to the state of river flow (i.e. it has rapids). The second shows one way in which we use water.

The inclusion of a person in each case helps the pictures to tell a story.

In each of these two photographs there were technical decisions to be made. In the second example the depth of field had to be sufficient to keep both the boy and the car in sharp focus. The exposure had to be precise enough to keep detail in the black car whilst also coping with the brightness of the sunlight reflected in the water.

In the end there had to be a compromise on exposure. The detail in the black was retained at the expense of letting some of the bright areas burn out. I am using this example because it illustrates the fact that much of photography is about balancing different elements - aperture and shutter speed, exposing for the highlights or for the shadows, going in close or retaining a wider view, and so on.

By taking a number of exposures we can try alternative approaches on the same subject. There are many instances, however, when we cannot achieve everything within one image.

Some of these issues have been met by developments in technology. For example, any desired combination of shutter speed and aperture can now be achieve either through modifying the ISO or by adding filters (e.g. a neutral density filter in order to reduce the dynamic range between solid black and blown-out white).

High dynamic range (HDR) techniques allow us to incorporate more detail across the highlight and shadow areas within a scene. Some of these approaches are discussed later. For now it is important to sort out the fundamentals.

It is impossible to over-emphasise that practice will help you to become more perfect. Experience, and the kind of guidance provided here, can help us to make the right choices for every photo opportunity that comes our way.

Lake Garda, Italy
Clouds, lake and mist - three forms of water.

First reactions to an image like the above is that there is not very much in it. That is almost the point. It is because of its simplicity that we can take it in and sense the atmosphere of the place and the moment. Quite often less is more. If in doubt about a composition just keep it simple.

The main purpose of this image is to show that water can occur in a variety of forms, in this case as a cloud, as water droplets forming a mist, and a body of lying water - the lake. This variety in water forms is the theme of the next chapter.

2

THE NATURE OF WATER

Steam, flowing water and hoar frost, three different forms of water, all within one scene. (Yellowstone National Park, USA)

Water can exist as:
1. snow and ice
2. hoar frost (frozen dew)
3. liquid water (streams, rivers, ponds, lakes and the sea)
4. water vapour (clouds, steam)

These forms may exist alone or in combination with each other. They may also occur in a variety of states (e.g. calm, turbulent, flowing, falling, breaking, splashing). Each brings its own photographic challenges.

However, as a starting point for photography, think of water as being either static or moving. This distinction is important as it influences the photographic techniques to be used.

When water occurs in its static form it can be photographed in ways that are similar to those used for landscape photography, only with some subtle differences (shown below and in *Section 8*)

Water that is moving requires your special attention (more of that later).

Let us look at some more examples of water types and states, and the best way to photograph them. The next two images are of sheets of calm water each providing a mirror-like reflection of the ground beyond.

The image above is of the banks of the **Lysefjorden in Norway**. The image below *(next page)* is of the banks of the **North Saskatchewan River, Alberta, Canada**.

NOTE: *In the Canada image a polarising filter has rendered the sky intensely blue. The fact that the blue is darker on the right and lighter on the left of the image is a usual feature of polarising filters. (This effect is related to the angle between that point on the scene and the position of the sun. More on polarising filters later.)*

Although these two scenes are at quite different scales, they require the same photographic approach.

Composition

With mirror images such us these the effect of the reflection is to double the size of the subject of the photograph. By placing the land-water boundary across the centre of the image you obtain 'two for the price of one'. With images like this be careful not to chop off the bottom of the reflection with the lower edge of the photograph.

Shutter Speed

Always make the shutter speed fast enough to avoid camera shake (if you must hand-hold). A good rule-of-thumb is to set the shutter speed to the reciprocal of the focal length of the lens (i.e. if the focal length of the lens is 100mm set a shutter speed no slower than 1/100th sec.).

Use a tripod whenever possible (whatever the shutter speed) and use the mirror lock-up facility if your camera has one. Whenever your camera is on a tripod turn off the vibration reduction (image stabiliser) setting on your lens.

Aperture

Close the aperture (higher *f*-number) until you get the required depth of field. (The normal term for the distance over which the image is acceptably sharp is 'depth of field'. However, my preference is for 'depth of focus' as this more specifically refers to a focussing property.)

Remember that any change in aperture will have an effect on the shutter speed. Double check that the aperture has not been closed so much that the shutter speed is now too slow to prevent camera shake.

Depth of focus

To obtain maximum sharpness within an image set the focal point as follows:
1. focus on infinity
2. examine the screen image and establish which subject nearest to the camera is still in acceptably sharp focus
3. change the focal point to fall on this subject.

The depth of focus should now extend from half-way between the new focal point and the camera all the way to the horizon.

Filters

Use a polarising filter whenever you need to counteract bright reflections from the surface of the water, and/or when you want to strengthen the blues and yellows within an image. Note, however, that a polarising filter reduces the amount of light entering the lens, and so affects the exposure setting.

Viewpoint

Both of the photographs (shown above, taken in Norway and Canada) were taken face-on in order to gain a symmetrical image with the subject and reflection making an equal contribution to the scene. However, be prepared to alter your viewpoint according to the nature of the subject and the kind of image you want to create.

For example, the next image was taken from a very high viewpoint (the top of a cliff) looking down on to a beach where a fresh water river is entering the sea. This provides a good habitat for seagulls. (Great Sur coast, California).

Within this picture three of the subjects are moving - the sea, the river, and the seagulls. Therefore a fast shutter speed (i.e. greater than 1/100th sec.) was required and so it was set at 1/250th sec. This automatically determined the aperture setting. A large depth of focus was not required because the river, sea and gulls are all roughly equidistant from the camera.

NOTE: *Once again a polarising filter was used - this time to cut out unwanted bright reflections from the surface of the water and to enhance the depth of colour in the sands.*

Photographically this last image also

shows how the near vertical view provides a picture of the dynamic relationships between river, sea, beach and birds. It also provides a good record of the spatial relationship between them. For example, some of the seagulls are wading in the fresh water, not the salt water.

The photograph also shows that there are interesting features (such as ripples within the river and cuspate edge forms within the sea) that would reward a closer treatment, but down at beach level, and from a lower angle of view.

Thirdly the image relies quite heavily on the angle of the light. It is low enough to throw shadows within the ripples of the river water and bright enough to form highlights on the surface of the sea.

Another form in which water occurs, is as lakes. The state of the lake surface may be calm and still, or rippled and rough - depending on the strength of the wind.

Few lakes, however, share the location or water quality of Crater Lake, Oregon. As its name implies, it lies in the centre of a volcanic crater.

Careful and very strict management has kept it free of pollution. Its dominant photogenic feature is the intense blue colour of its waters.

Crater Lake, Oregon, USA

The arrival of the two people dressed in white clothing was a bonus. It was possible to follow their movements until they came into the bottom right-hand corner of the image. This position gives the photo added interest and a sense of balance.

If we go now from this volcanic scene to that of Yellowstone National Park the opportunities for photographing water (in various forms) is very different.

Within the volcanic areas of the Yellowstone National Park water occurs not just as river water and snow, but also in the form of steam.

Yellowstone National Park (USA) *at the on-set of winter is one place where water can be seen in more than one state. Within this view snow, river and steam from geysers show water in three different states. The combination makes for a graphic scene.*

NOTE: *You will recognise (from the intense blue of the sky) that a polarising filter has been used - only in this case it has been overdone. The blue is too strong. So be warned - handle a polarising filter carefully.*

This Yellowstone image is really all about composition. The course of the river is used to lead the eye into the picture.

It is difficult to provide advice on composition. It is claimed that there are rules to follow - such as using a lead-in line (as here), placing a strong foreground object in the lower right quadrant of the image, providing a focal point to which the eye is drawn near the centre of the image, and so on.

There is a different rule which I find much more helpful. It is this:
'if it feels right it is right'.

To follow other people's way of doing things may be a good way to learn. However, it is not the way to become a photographer with a unique and creative spirit. An image that feels 'comfortable' usually contains a sense of balance. Now, back to the subject of water types.

The Great St Bernard Pass in Switzerland (the subject of the next image) has not got the steam of Yosemite, however the clouds more than make up for it. Water, snow and clouds combine with the glancing sunlight to provide another graphic scene of water in several states. This image is made more graphic by being almost a monochrome (black and white) though in fact it is a colour image.

Great St. Bernard Pass, Switzerland

NOTE: *A sense of calm has been brought to the surface of the lake (in contrast to the drama of the mountains and passing cloud) by* *using a slow shutter speed - it renders the water surface smooth and soft. This slow speed (1/4 sec) was used in conjunction with an aperture of f/22, this allowed maximum depth of field (focus), with the camera set on a tripod in order to prevent camera shake.*

TIP: *There is no need always to set the focus on the most distant point. Here it was set about two-thirds of the way into the image. With the small aperture of f/22 this allowed both background and foreground to remain acceptably sharp.*

Alpine images do not have to be sombre. On this day, (see image above) (Zermatt (Switzerland) located on the Matterhorn part of the Alps), the sun was shining, the air was clear, the sky was blue and it was tourist brochure kind of light. Alpine environments are normally associated with the presence of glaciers (e.g. the Swiss

Alps, Rocky Mountains, South Island of New Zealand). So down the left-hand side of the picture there is a glacier, with the curves on its surface indicating that it is flowing. The pattern of these curves shows that it is flowing more quickly in its centre than at its edges, where the ice is dragging against the ground.

Exposure can be tricky when an image contains both the extreme white of ice and snow and the dark areas of solid rock. The white of the snow, especially where sunshine is reflected from its surface, can burn out if the exposure allows for detail to be shown within the rocks.

HDR (high dynamic range) techniques have been developed in order to allow detail to be recorded throughout a final image. These techniques usually involve taking more than one exposure, with a different exposure setting for each frame. Enough frames are exposed to cover the full range of light and dark details.

These frames are then combined using software that selects the best exposed parts of each frame. In this way all details are recorded whether in a light or dark area, and the final image has a large (high) dynamic range.

In the example shown here (above) only one exposure has been used and it can only record solid black in the deepest shadows and pure white (without detail) in the areas of greatest reflected sunlight.

TIP: *Use your camera's histogram to check that the image records the maximum range of brightness values within any image of snow or ice.*

Svartisen Glacier, Norway.

This photograph of the Svartisen Glacier was taken from the deck of a cruise ship. We happened to be lying stationary within the fjord. As we looked at the scene a small local ferry was entering the view from the right. In order to give compositional balance to the image its exposure was delayed until the ferry was near the bottom right-hand corner of the frame.

The image was taken as a landscape subject (*with a combination of a 116mm focal length setting on a 80-200mm zoom telephoto lens, at 1/320th sec and f/8 using Fujichrome Provia Film rated at ISO400*). The focus was set at infinity as there were no foreground features close to the camera, and depth of focus was never going to be a problem. The relatively fast speed for a landscape shot was to counter any slight movement of the cruise ship.

This glacier is not only attractive but its form and the shape of the valley tells an interesting story about the glacier's behaviour, but more of that in *Section 13*.

At the other end of the scale frozen water can form smaller yet very photogenic forms. If we practice thinking of water at all scales from whole landscapes to single ice crystals, it greatly increases the potential for photographs.

Whatever the scale of the water image, as in so much of photography, the critical element to make or break the image is the nature of the light. The role of light is seldom more obvious than in the case of a rainbow.

Just for fun I have pumped up the saturation of the colours You will either love it or hate it. I think I hate it as, in general I am a fan of more muted colours.

The Hydrological Cycle

Scientists like to think of water as operating through a cycle known as the "Hydrological Cycle".

According to this way of thinking water is conceived as going round and round on a route that takes it from the clouds, into the rivers and lakes, and thence onwards into the sea. Once in the sea it evaporates and returns to the clouds, and the cycle starts all over again.

There are, of course, added factors such as the contribution made by subterranean water that may be seeing daylight for the first time. There is also the return of water into the atmosphere through transpiration in plants, and so on.

This is a concept that we, as photographers, can use. Our aim can be to photograph water at different stages within the hydrological cycle and recognise the stage that water has reached within the cycle as a whole. You will see what I mean as we go along.

It is time now, in each of the following Sections, to look more closely at the photography of water according to its particular form and state.

Exposure Readings and Water

Exposing 'correctly' for water can be very tricky. The light meter built-in to your camera will record the light intensity striking the sensor. It will take those readings from points pre-determined by the mode of light-readings that you have set. For example, a spot reading will be from one point (which you can set) and a matrix reading will be the average from a number of different points.

In general, built-in light meters do a very good job. However, the intensity of light as measured may not give you the exposure setting that you actually require. For example, you may want to keep all of the detail in the highlights within the water, in which case you may have to decrease the exposure indicated by the metering.

Exposure, therefore, becomes a matter of taste and preferences in the context of each image. So, don't slavishly follow your light meter, you, and only you, can decide on the 'correct' exposure.

One common practice is to take three exposures at a time. The first is your selected exposure, the second allows for some over-exposure and the third for some under-exposure. This allows you a safety net in case your first choice of exposure does not quite match your intentions.

This technique also provides you with three images to use if you want to make a new image based on all three exposure settings (i.e. using HDR techniques - see note on this at the end of the book). However, if you do this you should use a tripod to make sure that there is an identical frame and focus setting for each image.

3

CLOUDS

Much, but not all, of the water we see on earth comes from a cloud.

Very few photographers take photographs of a cloud unless it is part of a larger landscape scene. Why is that? Many clouds reflect or transmit light in an interesting way. They also have a form and texture that make them a worthwhile subject for photography.

Cumulus cloud over the Atlantic
(1/200th sec at f/11, 82mm focal length on 24-105 zoom, ISO 200)

The focal length of 82mm (on a zoom lens) was determined solely by the composition in order to show the cloud in the centre of the image and a horizontal trail of cloud going away on either side.

TIP: *When it comes to exposure be sure to check the histogram after each shot and see if an adjustment is needed.*

This cloudscape appealed just because of the interplay between light and shade, form and background.
(Exposed at 1/320th sec, f/5.6, 45mm focal length at ISO 640)

TIP: *A polarising filter is almost a standard accessory when photographing clouds. It helps to differentiate the clouds from the rest of the sky and brings out some of the detail within the body of the cloud.*

Clouds and light over Port Philip Bay, Melbourne (Australia)
(Exposure: 1/1000th sec at f/8, 105mm focal length, ISO 400)

Traditionally, however, photographs in which clouds may dominate also include their landscape context. The aim is to create the feeling that the scene is all about the form and the light produced by the clouds. To be more precise, it is all about the nature of the sunlight coming through or reflected from the clouds. Never is this more true than in the case of a sunset, especially if it takes place over a body of water.

Salt Lake, Utah, USA.

This image, though a landscape scene, is dominated by the (essentially cirrus) clouds. They are its main subject. They appear to radiate out from the centre of the image. Because the sun shines through them the the clouds provide the soft lighting that covers the whole scene. In a sense, therefore, this image is all about the nature of light rather than the character of the clouds.

The coincidence of cloud type and light within the landscape is very hard to predict. It is more a case of recognising it when you see it.

Similarly with the composition of an image that stems from the relationship between cloud and landscape. Sometimes the relationship between land and cloud is such that an image screams at you to be taken. The reaction of the photographer is normally spontaneous. Such was the case with the next image.

Cumulus cloud over Half Dome, Yosemite National Park.

Just for a few moments this cumulus cloud hung over Half-Dome, and then it drifted away. There was only one such opportunity in the whole day.

The moral of this story is:

'take it when you see it'.

If it gets better: just take another image.

Not all cloud photography is about fair-weather clouds. Indeed, some of the best cloud-dominated images are taken in foul weather.

Sometimes truly photogenic moments occur when shafts of sunlight break through. At other times such moments arise as clouds drift in and out amongst the hills and valleys.

Such interplay between clouds and land occur most frequently in mountainous areas. As illustrated by the following examples.

Hallstatt (Austria) with clouds mingling with the mountains.

As clouds come and go the amount of ground showing can change quickly and the best images will be available unpredictably and briefly. Timing the exposure can be tricky, but is easiest if a constant eye is kept on what is happening.

Photographically it may pay to hand-hold the camera using a shutter speed faster than the reciprocal of the focal length (the recipe for avoiding camera shake). (As a reminder: - if the focal length of your lens is 100mm you would use a shutter speed of 1/100th sec or faster). If your lens has built-in vibration reduction (or similar) so much the better.

By hand-holding your camera you are free to compose images quickly and catch each photo opportunity as it arises.

Here are three more examples. These are from The Canary Islands, Corsica and the Blue Mountains of Australia:

Mountain peak emerging from the clouds (Canary Islands).

Tree emerging from the clouds (Corsica)

Three Sisters, Blue Mountains (Australia)

Trees emerging from the clouds on a mountainside in Madeira.

In this example from Madeira the image is helped by the shaft of sunlight coming in from the top left, and the residual patch of cloud still hovering amongst the trees on the right. Conditions at the time were constantly changing. It was a case of having patience until the moment that the trees emerged from the drifting clouds.

These clouds, as in the two images that follow, were early morning clouds that were likely to clear as air temperature rose.

It was inevitable that lighting conditions would change rapidly. So much so that any two pictures taken 10 minutes apart would look different. In such situations it pays to linger and to keep photographing as the scene changes. The best images can be selected later.

Clouds and light (1), Blue Mts., Australia.
(Exposure: 1/50th sec., at f/9, with 70mm focal length lens, and ISO set at 200)

Clouds and light (2), Blue Mts., Australia.
(Exposure: 1200th sec at f/8 with 28mm focal length lens and ISO 200)

These two images show how lighting conditions and cloud form can change as the clouds lift.

It is clear from these few examples, that none of these images would be worth a second look were it not for the light that is playing on the clouds. A photogenic light is the essential ingredient for any successful photograph.

Just as the clouds come and go quickly so do the pools of light that fall on the landscape and create something special for us to photograph. There is, however, one form of dramatic lighting which lends itself very well to an 'atmospheric' mood in an image, and that is the dark menacing light of an impending storm.

Storm gathering over the Himalayan mountains above Dhahran, Nepal.

Impending storm, Scottish Highlands, UK.

The darker areas in the lower image have been allowed to go fully black. This works because of the overall sombre and moody nature of the images. It also enabled detail to be retained within the brighter areas of the sky, and especially within the shaft of light coming through to form a single spot of sunshine on the landscape.

This control was achieved by under-exposing the image by one *f*-stop.

Sometimes, as in the next picture it is possible to use these sombre conditions to tell a dynamic story.

Storm over the lake at Koningsee (Germany)

This picture of Koningsee gives a strong sense of the link between cloud and lake. It is clear that the precipitation of the storm will contribute to the amount of water in the lake.

It graphically illustrates one of the important components of the hydrological cycle. It shows the direct supply of water from cloud to lake. From here it will progress along the river system to the sea.

A different relationship between clouds and valleys is created by an inversion of temperature. This tends to happen in the winter when cold air can sink into the valley bottoms. If there is enough moisture this will condense in the cold air and form clouds at low levels. When this happens the best position for photography is on the valley side above the clouds.

Clouds in the valley of the River Derwent, Derbyshire (UK) - formed as a result of temperature inversion.
(Exposure: 1/50th sec. at f/5.6. 200mm setting on 70-200mm zoom, and ISO 800)

To close this section on clouds it is appropriate just to linger for a moment with a sunset (or two). The first of these was taken in Corsica, the second at Lake Tahoe, California, USA.

During a sunset light intensity is inevitably low, and care has to be taken over keeping the camera steady during the exposure. This is achieved either through using a high ISO value, or by using a tripod. Plump for the second option if you possibly can.

Set the tripod up so that you have the composition you want. Sort out the depth of focus that you require - this will be dependent on how close the nearest foreground object is to your camera (this must be in sharp focus).

Sunsets can provide vivid and photographically irresistible moments.

Leave the camera set on automatic, make sure image stabilisation is OFF (as it always should be when you are using a tripod), and use a shutter release cable (or remote shutter firing device) to take an image every time the sunset changes to the next spectacular display of colour. Wait to see the sun drop below the horizon. The best sunsets tend to occur about 10 minutes after the sun has gone down.

Lake Tahoe, California

However, always be prepared to break the 'rules' if there is a good picture in front of you. The Lake Tahoe moment was there to be taken. The sun had not yet set below the horizon and the light was fast going.

If you look carefully you will see that I have hidden the full brightness of the sun itself behind the trunk of a tree. This left a range of colours and light intensities that could be handled by the sensor without the sun's highlights burning out.

TIP: *follow the guidance until there is a good reason not to do so.*

4

DEW AND RAIN

Clouds lead to both rainfall and condensation. This may leave raindrops hanging from twigs and branches, providing an opportunity for macro-photography.

One of the stimulating things about recording water in its various forms is that it takes us from distant clouds at one end of the scale to recording single drops of water at the other end of the scale.

As photographers we must be able to adapt our vision, scale of observation and technique to fit the opportunities that water provides.

In this picture (of the water droplets hanging from a branch) it was important not to have a distracting background. I moved around until I had a deep shadow behind the droplets.

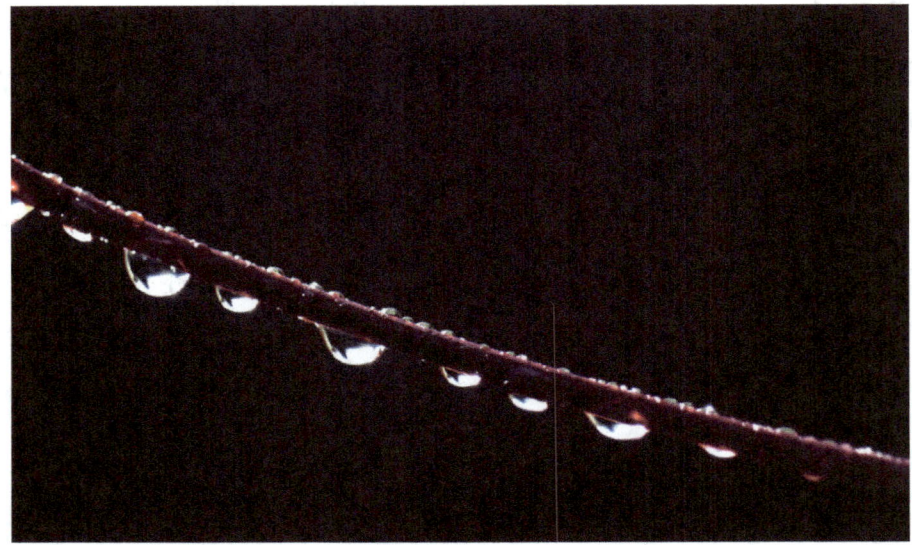

This image of water droplets required a tripod in order to do two things:
1. keep the composition set in a fixed place
2. prevent inadvertent backwards and forwards movement of the camera (which would have jeopardised the sharpness of the image).

Levels' (in Photoshop) was used in order to turn the shadow into a full black- (i.e. using 'Levels' the left-hand slider was moved into the histogram until the dark area was fully captured as black).

In the next image low cloud and early morning dew has settled on a eucalyptus forest in the **Blue Mountains, Australia**.

(Exposure: 1/15th sec at f/8, focal length 47mm and ISO 400)

Several photographic questions arose in the photographing of this image:

1 Where to focus? I went for sharp focus on the foreground rocks. The mists destroyed any sharp definition of the tree trunks in the distance, so there was no point in trying to make them sharp and loose definition on the clearer objects nearer to the camera.

2 How to handle the exposure? It was important not to loose the feeling of the low clouds, mist and dew on the ground. I pulled the exposure back a stop (i.e. I decreased the exposure) as I did not want the mist to burn out. I then checked the exposure histogram (on the camera) to make sure that no data was being lost at the bright end (right-hand side) of the scale.

3.   How to process the image? Simple, I stayed away from any sharpening of the image.

The result was this soft rendering of a magical scene.

A heavy impending storm, however, yields an image in sharp contrast to the softness of the eucalyptus forest (above).

Storm clouds - opportunities for photography! But, take care of your camera! Taking an image in the pouring rain can present its own challenges (e.g. keeping the lens free of water droplets). However, it can also yield images that have some unique properties.

Storm cloud over Funchal, Madeira

The next two images were taken in the pouring rain on Lake Koningsee (Germany). In both cases the colours of the damp buildings have been enhanced by the fact that they are wet.

In the first image you can also see the splashes of rain on the surface of the lake. In the second example remnants of the rain cloud appear in the image. Both of these elements contribute to the nature of these images.

Koningsee, Germany – in the rain

So, the fact that it is raining is not a good reason (by itself) to put your camera away. However, it is a wise precaution to protect the camera - either by asking someone else to hold an umbrella over it, or to put it in a purpose-made protective housing (usually a waterproof bag with a hole at the front for the lens).

Now for something different. You will come across photographs that have been taken through the windscreen of a car when it is raining. The effect is to distort the forms and colours of the objects outside.

Here is a variation on that theme. Use has been made of water running down a window of an art gallery in Melbourne (Australia). The rain running down the window created a distorting effect on people passing along the street outside.

This mother, child and pushchair hurrying past was caught in an instant. The whole process of taking this picture was pure photographic instinct. The camera was on auto everything. Fortunately the aperture was set at f/11 (just enough for the required depth of focus). The auto-focus picked up the glass of the window because there was enough water running down it to make the glass appear as a tangible object on which the auto-focus could register. Before I could change anything the mother, child and pushchair had gone.

We sometimes strike lucky. But, we cannot be lucky if we are not there with our camera and on the lookout for every opportunity that comes our way - even when it is raining.

The picture may be in the general scene, or it may be in the detail. The final picture in this section was taken in a road-side lay-bye.

Jewels of water droplets that have been retained on a cobweb. The picture is in the detail.

5

FLOWING WATER

River in Oregon, USA

Once the rain has gathered into streams water begins to flow, and in doing so it presents us with a whole new challenge. We have to deal with moving water.

Movement can be fast, or it can be slow. It can occur as a continuous sheet of water, or it can be as a waterfall. Flowing water should cause us to think carefully about how we are going to record it. This applies not only to the fact that it is moving, but also to how much of the water we want to include. Do we include its setting, do we just photograph the river banks and the channel, or do we go right in and photograph the detail of the flow itself?

The mind has to be like a zoom lens when it is searching for a good picture. It has to be prepared to view its subject either as whole (the wider view) or in part (the telephoto view).

We have to be able to perceive the best picture at whatever scale it is to be found. This is where personal preferences come to play. Some will like the broader view (as in landscape photography) others will want to go in tight. For me the best images are often in the detail rather than in the broader view. See what you think as we go through some examples.

The focal length that I prefer most for river flow images normally falls in the range 70-135mm. However, the final choice also depends on the size of the river and the viewpoint.

Sometimes it pays to review your old images just to see what could have been done better. When I did this for some river images recently it was clear that there were pictures within the views that I had captured, worthwhile pictures, but they were not the ones that I had taken. Some of these were pattern pictures, others could have encapsulated in one image the power of the flowing water. These were opportunities that had been missed. That review of past images made me more determined to spend time on riverbanks looking for small elements of river flow that could make a pleasing composition. The results were much more than that. Some are included below.

Patterns in rivers and streams are created by three elements. These are the flow of the water, the disposition of rocks (if there are any), and the nature of water splash. Concentrating on the interplay between these three components provides the essential elements of a worthwhile image.

There is always the potential for an infinite combination of the three components: water flow, rocks and splash, each one yielding its own unique image. In general, fast flowing streams in upland areas provide more photographic opportunities involving these three elements than a wide river flowing through lowlands.

There are two other important factors in the photography of flowing water. These are the nature of the light and the length of exposure.

Light may be reflected from or transmitted through water. It can bring extra life to droplets in splashing water.

One of the best locations for images of river detail is within a section of rapids, for it is here that turbulence sets in and rocks start to be exposed through the water.

Light on rapids, Scotland, UK

Padley Gorge, Sheffield, England.

This picture sets the general scene, but look carefully at the flow of the water and you will see many opportunities for images of the river as it flows.

The next two images illustrate this point.

NOTE: *In the case of every photograph involving flowing water decisions have to be made concerning shutter speed.*

In this image the shutter was open for 2 sec whilst that in the next image it was 1/5th sec.

So, do you favour a slow shutter speed that turns the water more milky (the first of these two images), or would you have made it faster and frozen the water in its tracks?

TIP: *Take several i*mages at different shutter speeds. You can decide later which you prefer.

The 'best' shutter speed for recording flowing water, has long been a matter for debate. Some photographers like their water to look diffuse, and hence use a long exposure (i.e. longer than 1/10th sec). Others like to record the droplets as clearly defined elements in the image, and so go for a short exposure using speeds of 1/100th sec. or faster. In either case a tripod is indispensable as an aid to composition.

As a scientist who has spent more than 30 years studying river behaviour I find images taken with a long exposure to be very revealing.

By allowing the movement of water to blur the image the

photograph shows where the greatest amount of energy and power lies within the flowing water.

Where the water is particularly turbulent most of the erosion of the river bed is likely to occur. If this turbulence is captured in close-up the image seems to adopt some of this power.

Turbulent flow causes the river to erode its bed and its margins – the river banks. It also allows a river to carry this eroded sediment further downstream, as will be discussed later.

(previous page)
A cameo shot of river flow using slow shutter speed (1/10th sec) and portrait format
(determined by the orientation of the subject).
The slow shutter speed reveals the locations of the most powerful forces within the water.

The direction and concentration of river flow can be detected more easily when a longer shutter speed is used.

Note: *Bright sunlight reflecting from water surfaces can produce difficulties.*

For example, bright reflections can make the calculation of the correct exposure tricky. Remember, not to expose for the brightest areas. Instead, take an exposure reading from a mid-tone area. Now take three shots (i.e. bracket your images) by taking one at the exposure you have recorded, another with the exposure increased by

+0.3EV and another at an exposure of -0.3EV. (If you are unsure what this means consult your camera handbook or ask a more experienced friend).

Another approach is to use a rotating circular polarising filter to tone down the unwanted bright reflections.

When rivers and streams are long and thin it may be worth using a portrait rather than a landscape format. (Canadian Rockies) This is especially true if you are looking straight up or down the length of a narrow river.

When a river is wider a landscape format allows more of the banks and surrounding scenery to be included.

(Cheedale, Derbyshire, England)
(Exposure: 1/30th sec at f/11, focal length of lens 80mm, ISO 400)

The best lighting conditions; with far fewer exposure difficulties occur with a bright but over-cast sky rather than direct sunlight.

By concentrating your exposure on a small portion of a river surface the image can almost take on an abstract form. It becomes a 'pattern picture'.

Surface form of the River Teign, Devon, England

6

FALLING WATER

Water falling over a man-made weir.
(Monsal Dale, Derbyshire, England)
(Exposure: 1/80th sec at f/11 lens focal length 80mm ISO 400)

As rapids become higher they soon qualify as waterfalls. Much of the advice already given on the photography of rapids also applies to the photography of waterfalls.

This Section provides examples of waterfalls and comments on the photographic techniques used. You can use these ideas as a basis for your own approach to the photography of falling water until you think of other ways of doing things.

It is a matter of taste as to whether or not a foreground feature should be included when photographing waterfalls. It may depend on what is available at that location.

For the Mackenzie Falls (below) several shots were taken. The one shown here allowed a branch to come between the camera and the falls. The effect is to give a sense of depth to the scene that otherwise would have been missing. It almost creates a 3-D effect.

Mackenzie Falls, Grampian Mountains, Victoria, Australia

There are times (as here) when the environmental setting provides a valuable context for showing the waterfall.

Moss Glen Falls, Granville, Vermont

Falls in Glen Nevis, Scotland
(Exposure: 1/3rd sec at f/22 with 165mm lens ISO 100)

These falls in Glen Nevis drop under a road bridge (of no great architectural or pictorial merit), so the bridge was excluded from this view. A 165mm lens (with the camera mounted on a tripod) was used to obtain this composition. This lens made it possible to omit everything except this one section of the waterfall.

Distracting features should be left out. The one place where they tend to creep in is around the margins of the image. Before pressing the shutter always check the image margins and make sure that no unwanted objects have found their way into the picture.

One way of altering the content within the frame is to step forwards or backwards. This is the only effective way when you are using a fixed focal length lens.

If you are using a variable focus lens you can, of course, zoom in or out of the picture in order to include less or more of the scene in front of you.

This was the case with the next image, where a zoom lens had to be used because there was no room to move any further forward..

Detail, Moss Glen Falls, Granville, Vermont.
NOTE: *This and the next image are all about shutter speed (slow in this case) and the way that light from above is reflected from the surface of the water.*

***Base of waterfall into Milford Sound,
South Island, New Zealand.***
(Exposure: 1/30th sec at f/9 lens focal length 90mm, ISO 400)

In this case only a small part of the waterfall has been included. This part was selected because of the spray, the ripples radiating out across the surface of the water, and the nature of the light emanating from the water droplets within the fall. This photograph was taken from a boat. It formed a very unsteady platform. However, by timing the shutter release with the moment that the boat was between rising and falling several shots survived the experience.

The timing of the visit and the viewpoint were entirely out of my control. But, as so often is the case in photography you take the chances as they present themselves. The two images that follow were taken at the same time as the previous image. The first is of a higher part of the same fall and is shown as taken.

Normal exposure.
(Exposure: 1/40th sec at f/6.3. lens focal length 120mm, ISO 400)

The second is of the base again, but this time the image has been processed so that the rocks have been rendered completely black allowing the water to stand out against this dark background.

Digitally processed to create a graphic representation of the base of the waterfall.
(Exposure: 1/30th sec at f/9, lens focal length 90mm, ISO 400)

7
BREAKING WAVES

Breaking waves present the same challenges concerning moving water as waterfalls. However, there is one critical difference. Most photographers appear to like to catch the breaking wave when it displays maximum impact. In the days of film photography we had to wait until we had a sense of the timing between waves, and then hope to press the shutter at just the right moment.

Those days have gone. Now the answer is to set the camera on continuous shooting, and then pick out the winner. That is the easy bit.

Wave breaking against rocks, Devon, England

Heavy swell Seven Sisters, Sussex, England
(Exposure: 1/500th sec at f/11, lens focal length 150mm, ISO 400)

Breaking waves on the shore are caused by a combination of strong on-shore winds and shallow depth of water.

If the camera is turned towards the shore where the shallows occur, the drag on the base of the wave causes it to break and to transfer its energy to the beach.

When this process take place at the base of a cliff it causes an undercutting of the cliff face and may bring about rock fall or landslips within the cliff above.

Waves breaking on the shore below the Seven Sisters, Sussex, England *(Exposure: 1/800th sec at f/11, lens focal length 140mm, ISO 400)*

The steepness of the chalk cliffs seen in the background is caused by the constant erosion of its base by powerful breaking waves such as those captured in this image.

This is water at work in sculpturing the form of the cliffs. It is nature at work (see *Section 13*).

Breaking Waves, Johanna Beach, Great Ocean Road, Victoria, Australia
(Exposure: 1/125th sec at f/13. lens focal length 105mm, ISO 400)

With a wider view of a stretch of beach timing is easier and there is no practical need to set the camera on "continuous shooting" mode.

A completely different set of circumstances occurs when you are out at sea. The breaking waves around you can create fascinating forms, as here in the Bay of Biscay (next page).

The problem was that we were in a Force 12 gale, and the ship was in constant motion.

The trick was to combine a fast shutter speed (1/1000th sec) with the moment when the ship was between going up and going down.

The practical alternative is, once again, to set the camera on continuous shooting.

Waves breaking below sea cliff (South Australia)
(Exposure: 1/80th sec. at f/22, lens focal length 105mm, ISO 200)

A diffuse glancing side-light always adds that something special to any image.Breaking waves tend to have a bright top edge and bright spray. These can cause exposure problems. There is a tendency for such bright areas to be over-exposed and loose detail and definition.

When this happens decrease the exposure by one *f*-stop (more if necessary). It doesn't matter if the darker areas turn black (as in the picture above). At the image processing stage try the following: decrease the exposure and then increase the brightness until the image contains a full range of tones.

Power in the waves. (Sidmouth, Devon, England)
(Exposure: 1/100th sec. at f/16, Lens focal length 80mm, ISO 100)

Spray shows up best when there is a contrasting background.

Never forget how dangerous the coastal environment can be. So don't be one of those photographers who is tempted to go too near to the action. Many cameras have died this way, and so have a few photographers.

8

STILL AND CALM

Dawn over the River Seine, France

Now the mood has changed completely. But has it affected the techniques we use to photograph water? There is more time to contemplate water when it is still and calm. This state implies "peace" and our photographs need to pick that up. One way to do that is to reveal the stillness of the surface by having a mirror image reflection.

Early morning, Ardennes, Belgium.
(Exposure: 1/400th sec at f/9, lens focal length 45mm, ISO 400, tripod)

Not only is there a reflection but there is also an early morning mist hanging over the water. This implies still air, and further enhances the message of peacefulness.

It is a mistake to think that such images always need to be full of colour. In many ways a more subdued almost monochromatic representation gives a calmer feeling than an image full of vibrant colours.

This is the case for the next image (as well as the one above).

Wallaga Lake, Bermagui, NSW, Australia
(Exposure: 1/30th sec at f/11, lens aperture 50mm, ISO 200, tripod)

Note: In both of these two examples (and several of those that follow), where the reflection adds so much to the image, the horizon has been placed across the centre of the scene. When the strip of land is so thin the whole image is about sky and its reflection in the water. Keeping the proportion of sky and water equal becomes an important element in the success of such an image.

Reflection in still water, Dinant, Belgium
(Exposure: 1/30th sec at f/11, lens focal length 24mm, ISO 200, tripod)

The inclusion of colour is essential here because of the subject of the photograph (i.e. the internal lights within a building in the twilight).

That is not crucial here. What really matters is that the sky still retains a blue colour. It is usually much better to take an evening image <u>*before the sky turns black.*</u>

Barricane Beach, Devon, England (taken with camera mounted on a tripod).

A tripod allows (i) a small aperture and slow shutter speed (thereby gaining maximum depth of focus), and it enables (ii) precise composition with a horizontal skyline and a hint of rocks on the left-hand side (these hold the eye within the picture and give it a better sense of balance).

Still and calm conditions allow the use of a tripod to maximum effect.

Exposure times can be long (i.e. in excess of 30 secs.). Composition has to be precise (e.g. when making sure that a shoreline is placed precisely across the centre of the image), and use can be made of low light either early in the morning or into the evening.

The tripod needs to be sufficiently stable to do the job. A cable release (or remote equivalent) will help to prevent camera shake. If your camera allows you to lock up the mirror then do so).

(**Note**: *not all cameras have mirrors that need to be locked out of the way!*)

Circumstances can occur that are less in our control.

Whilst travelling on the River Danube the water surface was exceptionally calm. The deck was busy with people so there was no point in even attempting to use a tripod.

The solution lay in turning up the ISO setting on the camera. This allowed a faster shutter speed.

Water surface, River Danube, under calm conditions.

The ISO was set at 2500, the shutter speed was 1/250th sec, and the aperture was set wide at f/5.6 (depth of field was not a limiting factor for this subject).

A similar situation arose later in the same day when the sunset and its reflection across the calm waters of the Danube led to the same technique being used.

Under normal conditions a tripod is absolutely essential during night-time photography.

Eastbourne Pier (UK)
(Exposure: 10 secs at f/22, focal length 84mm)

There follow a few more examples of water in a still and calm mode, just to illustrate what is possible.

Approaching Stavanger (Norway) in the early morning. Calm but not a smooth surface.

This dawn picture was taken from the deck of a cruise ship. The vessel was moving very slowly, but even so there was a perceptible motion that made it difficult to take a sharp image without camera shake. Using a tripod would not have made that situation any better.

The only alternative was to use a high ISO and a shutter speed fast enough to overcome any such movement. Such a faster shutter speed meant that no advantage could be taken of the smoothness that a slow shutter speed would have introduced.

By having to use a faster shutter speed (1/1000th sec) the image is able to record the micro-ripples set within the tiny waves that are dominant to the eye. This reveals another level of dynamic information about the moving water.

Backwater, Merimbula, Australia. Calm and smooth.
(Exposure: 1/100th sec at f/8, lens focal length 28mm, ISO 400)

Stillness brings with it reflections of the land beyond and the sky above.

Reflections are always darker than the subject that is reflected. The trick is to expose for the subject above the water and let the reflection look after itself, as in the next images.

Wollaton Hall, Nottingham, UK - across the man-made lake early winter morning.

Note: When creating a mirror image like this a successful composition usually comes by placing the land-water boundary halfway up the frame.

Moritzburg Castle, Saxony, Germany Luxembourg City.

***Lake emerging from the clouds,
Nr. Tamworth, New Hampshire, USA***

Lake and sky, Whittington, Vermont, USA

Still and calm waters are amongst the most photogenic landscapes that you could ever wish to photograph. Just how much this is the case is a function of the nature of the reflections from the surface of the water. In this the role of the sky is often crucial.

However, you will notice that in the Whittington image the detail of the reflection is broken up by the vegetation floating on the lake surface. The role of the water surface itself is also important to the final image.

9

SALINE LAKES

Mineral stack, Mono Lake, California, USA

Although saline lakes form a very particular kind of environment for wildlife and vegetation, they require the same approach (in terms of photography) as the still and calm situations described in the previous section. Their salinity is due to the rate of water loss through evaporation in the high temperatures. The amount of loss through evaporation is greater than the supply of fresh water from the surrounding hills.

The saline lake featured here is Mono Lake, California. In the case of Mono Lake there has also been a decrease in the volume of inflowing fresh water because of the diversion of these tributary streams for use in the coastal towns and cities of California.

Over the years continuing evaporation (without the renewed supply of fresh water) has caused the surface of the lake to drop several meters, and its salinity to increase. This has left stacks of minerals exposed that were originally created underwater in those places where fresh water interacted with saline water, causing these minerals to be precipitated into solid forms.

The drop in lake level that resulted from the diversion of the fresh water has now been halted, and in time it will rise again to submerge the mineral stacks. This reversal in fortune is the result of the prohibition of further water diversion. Our current view of Mono Lake is therefore destined to be a temporary one.

The margins of Mono Lake, as with many other salt lakes, are the home of a massive population of flies. These in turn are the food supply used by migratory birds. This makes such a lake a good place for bird and wildlife photography.

Migratory birds feeding in the Morning light, Mono Lake, California, USA

Rippled reflection of saline stacks, Mono Lake, California, USA

Mono Lake is only one of many saline lakes around the world. All of these form a special environment for wild life, but few (if any) others have these remarkably photogenic mineral stacks.

Salt lakes, such as that in Death Valley (which come and go with the rains) do not have these stacks simply because the same underwater formation conditions do not exist. Instead, when they dry up they leave an extensive plain of surface salts, precipitated as the water evaporates.

Dry bed of saline lake within the Basin and Range province of the USA.

10

STEAM

Steam vents, Yellowstone National Park, U.S.A.

There are a number of volcanic sites around the world where steam is emitted. The following images were taken either in Yellowstone National Park, or at Rotorua on the North Island of New Zealand.

These vents are usually set amongst hot springs, mineral pools, former volcanic craters and calderas.

Photographically we can handle these in a number of alternative ways. Most people will take their holiday snap of the geyser erupting and think "job done".

However, there is more to it than that. Geysers can be taken in their general setting, or they can be taken in close-up, with the image concentrating on the base of the geyser where the steam is emitting from the ground.

Steam by itself doesn't form much of a picture, but place a solid stable feature in the photo as well and you have the potential for a good image.

When taking the steam set against the sky wait for the sky to have a contrasting colour (e.g. bright blue or dark grey). With a mid-grey sky the steam will merge with the sky. Like the spray from a breaking wave, it needs to be well defined by a colour contrast.

Geyser, Rotorua, North Island, New Zealand
(Exposure: 1/250th sec at f/8, lens focal length 105mm, ISO 200)

*Steam rising from a vent,
Yellowstone National Park, U.S.A.*

Hot pool in its setting, Yellowstone.

Mineral Pool (1),
Yellowstone National Park, USA

With such pools it helps to have a tripod in order to allow good composition, a long-focus lens, and a slow shutter speed.

It is impossible to use the photograph to judge the size of the pool.

You can also to treat these colourful images as abstract forms of nature. They could be used as such in a gallery or in a photographic competition.

*Mineral Pool (2),
Yellowstone National Park, USA*

In Yellowstone mineral pools can be found where there is a feature by which we can judge its size - in the case of the next illustration it is a dead tree.

Dead tree amongst the volcanic features.

Not all sources of steam are natural.

Puffing Billy, Victoria, Australia

In pictures like this, where the steam dominates, it is important to retain something of the locomotive so that the viewer is able to tell what is going on.

Steam combined with snow, hoar frost and running water within one scene, Yellowstone National Park, USA.

It may seem perverse, as some of the previous images have shown, that steam can co-exist with snow as it does in Yellowstone National Park in the winter. The steam comes from hot conditions deep underground whilst the snow is a result of the cold temperatures in the atmosphere.

Don't knock it - these conditions provide great opportunities for the photographer.

11

FROST, ICE AND SNOW

Fritz Range from Franz Josef Village, South island, New Zealand.
(Exposure: 1/100th sec at f/6.3, lens focal length 170mm, ISO 400)

Alpine environments, whether the Alps in Europe, the Andes in South America, or (as here) the mountain ranges of South Island New Zealand, are always a hot favourite with photographers. Not only do they provide dramatic landscapes with opportunities for photographing ice and snow, they are also associated with crisp clear air. These make for sharp images of grand and beautiful features, with ample variety and many sites where some of our best landscape photographs can be created. Nearly all of this is the result of the presence of water (in one form or another) or the result of work carried out by both ice and running water (see *Section 13*).

The Alps, Grindelwald, Switzerland
(Exposure: 1/100th sec at f/11, ISO 100)

Alpine environments also provide a chance for creating images from a high viewpoint, whether from the ground or an aircraft.

Ice cap above the Franz Josef Glacier, South Island New Zealand.
(Exposure: 1/1000th sec at f/ 8, lens focal length 70mm, ISO 400, taken from helicopter)

The image above was exposed at 1/1000th sec from a moving and vibrating helicopter. The choice of shutter speed was a guess. Fortunately it paid off and the resulting image was sharp. An alternative would have been to increased the ISO just short of the point when quality drops away. This ISO value varies from camera to camera.

The helicopter trip had been booked for early in the day so that the light would be soft and glancing. Harsh shadows would have killed this image.

When ice lies in a layer like this it does much less erosive work than it does when it is moving. The ice here is forming a protective blanket over the slopes below. Once it becomes a glacier that is no longer the case and erosion dominates (see *Section 13*).

In the next image the ice can be seen breaking away where the ground underneath is steep enough for the ice to move under gravity. This causes the distress that we see at the surface.

This is a good example of where the forms we see, and photograph, betray the processes that are happening.

In the case of ice it is often the case that a smooth surface to the ice suggests that it is protecting the ground underneath.

Conversely, broken ice at the surface, especially if it has a downhill gradient, implies that erosion is taking place at the base and the sides of the ice.

***Disintegration on edge of France Josef Ice cap,
South Island, New Zealand***
*(Exposure: 1/1000th sec at f/8, lens focal length 70mm, ISO 400,
taken from helicopter)*

***Crevasses and clouds, Franz Josef Glacier,
South Island, New Zealand.***
*(Exposure: 1/800th sec f/8, lens focal length 97mm, ISO 400,
taken from helicopter)*

The crevasses in the surface of the glacier show that it is moving. More on this in *Section 13*.

The whole image owes much to the morning sunlight that is beginning to break through the clouds .

Lower down the mountain the ice turns into the valley of the Franz Josef Glacier. Its broken surface indicates that flow has speeded up and the whole ice body is heading down the valley, eroding as it goes.

Head of the Franz Josef Glacier
(Exposure: 1/1000th sec at f/8, lens focal length 85mm. ISO 400)

Water in its solid and most dramatic and extensive state occurs as ice caps and glaciers. The most recent glaciations (the *Alpine Glaciation*) are well past their maximum, though remnants of these glaciers still remain in the higher mountains, such as the Alps and the Himalayas.

As they melted these glaciers left behind significant traces of their former extent, including corries, U-shaped valleys, and moraine.

Where moraine was left across the width of the glacial valley (a terminal moraine) it can form a dam behind which lakes still remain. Such a lake lies in the Canadian Rockies, and is appropriately named 'Moraine Lake'.

Moraine Lake, Canada

At the other end of the scale frozen water can occur as snow flakes that may cover the whole landscape. This is a time when photographic opportunities abound.

Once again these are found at a variety of scales from the very general landscape view through to images of snow crystals. Starting with a general landscape view:

View towards Kinderscout, Peak District National Park, UK
(Exposure: 1/125th sec at f/10, lens focal length 138mm, ISO 200)
In terms of composition, notice how use has been made of the tree in the bottom right-hand corner.

On the plateau tops in this part of the Peak District National Park, the winter landscape casts a bleak grip across the summits.

Over-night snow fall has transformed the landscape. (Massif Central, France)

When blown by the wind drifting snow can be a hazard to property. (Massif Central, France)

Even without drifting it can make life more difficult. Dore, South Yorkshire, England
(Exposure: 1/50th sec at f/22, lens focal length 58mm, ISO 400)

Although you can always try to let life go on as normal!

Staying true to the idea that as photographers we should always be looking for images at every scale, there are many photographic cameos to be found within the details of a winter scene.

Shadows and in the snow.
(Exposure: 1/160th sec at f/9, lens focal length 187mm, ISO 200)

When photographing snow the one thing that requires special care is exposure . Make sure that you check the exposure histogram on your camera after every shot.

If this histogram shows that some of the light values are falling off at either end of the data range it will be necessary to modify the exposure.

It is commonly the case, in snow, that you need to add at least one *f*-stop to the reading that the exposure meter gives you. The meter will 'see' white snow as mid-grey in colour. An increase in exposure (above that indicated by the meter) will help to correct this.

There is a good case for bracketing your exposures when there are large areas of reflective snow and/or ice around. Then you can choose which is the best at a later stage. Snow is, in effect, frozen water. Frozen water can take on a number of other forms such as ice crystals and icicles. Some of the most photogenic forms can occur as ice on pools of water (below).

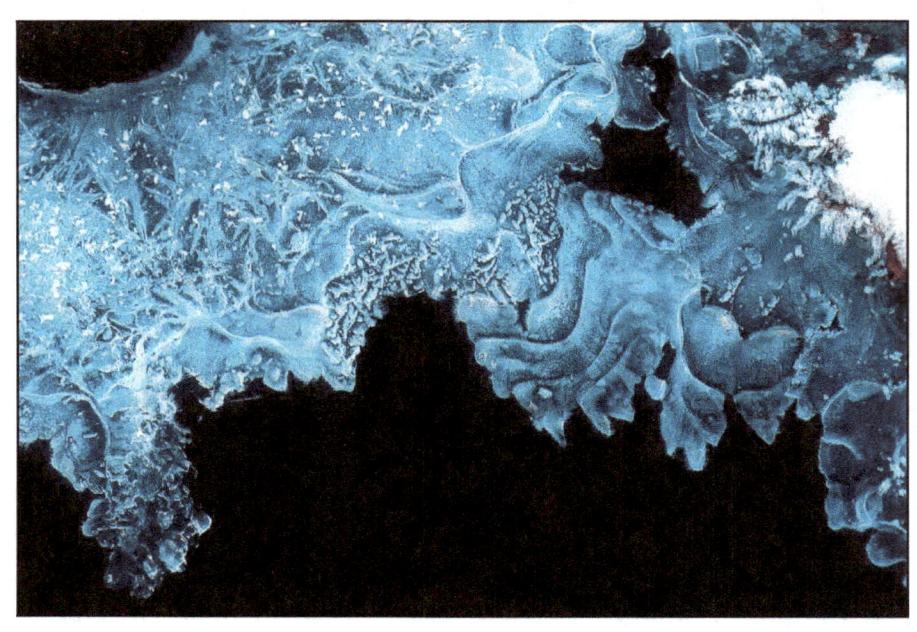

Icicles formed by splashing water on the margin of a stream.

Icicles formed on twigs as a result of water dripping from above.
(Exposure: 1/100th sec at f/8. lens focal length 105mm, ISO 800)

Icicles below a roof gutter, set against the sky
(Exposure: 1/200th sec at f/5.6, lens focal length 105mm, ISO 200).

Snow crystals

Photographing at this very detailed scale we are in the realms of macro-photography. The essentials for achieving success include having a lens with a macro facility (if not a macro-lens), a tripod, and skill in close focusing.

Hoar frost on trees

12

WATER IN PICTORIAL IMAGES

*Graphic image of trees in snow,
Chatsworth Park, Derbyshire, UK
(Exposure: 1/320th sec at f/16)*

Many of the images in the earlier Sections could be described as pictorial. Indeed, some of them were originally taken because of their pictorial nature.

That, in many ways is the point of photographing water, it allows strong pictorial images to be created.

Both of the first two images in this Section were taken entirely for their pictorial value rather than just because they contain water. However, it is the water component that makes them worthwhile.

Reeds in lake, with sky reflection, Isle of Skye, Scotland.

The same can be said for the following images. They have water as the dominant component, and yet they are essentially pictorial.

River Cam and boat house, Audley End (UK)
(Exposure: 1/160th sec., at f/16. Lens focal length 70mm, ISO 400)

Porto Bay, Corsica

Alde Estuary, Suffolk, UK

River joins the sea, Petit Bot Bay, Guernsey, Channel Islands.
(*This image was created as a pictorial scene. Note how the line of the water flowing down the beach is a useful lead into the more distant part of the image where the children are playing.*)

For this next section I am staying on the beach at Petit Bot Bay in order to illustrate some of the pictorial images that can be created within one small area.

Recently, and quite by accident, I made a discovery. None of my photographer friends can sit still on a beach.

This was a great comfort as I had always been under the impression that I was alone in my desire to get up and take a good look around. Maybe people thought I was looking for shells, or crabs, or a lost contact lens. In truth I am always studying beach patterns in search of images.

Amongst these my favourites are those that show the effects of water flow (usually fresh water from the land) across a beach. The resulting photograph can be either pictorial or, in some cases, even abstract in nature.

The best beaches for such pictures are those with a significant amount of sand preferably enclosed between two rocky headlands. The examples in this Section come from Petit Bot Bay, only a mile from the airport on the south coast of Guernsey (Channel Islands). This bay lies between two granite headlands. It is largely composed of sand, with perhaps twenty per cent of the beach covered in rocks and boulders. The tide rises almost to the foot of the cliffs, but when it recedes a stretch of beach over one hundred yards wide is exposed.

Across this beach there flow two streams of fresh water. The smaller one emerges from the cliffs on the western side of the beach. The other enters from a valley at the head of the bay, and has the larger volume of water. Initially this larger stream crosses a bank of boulders thrown up during storms, or fallen from the cliffs. Some of these are covered in seaweed. The lower part of this boulder beach receives fresh sand during high tides. When the tide retreats the fresh water stream washes much of the sand away.

Beyond and seaward of the boulders there is a continuous cover of sand. It is within this area that the stream divides into a number of separate channels. Here the water flow is shallow compared with higher up the beach, but it occupies a much wider zone now that its course is no longer constrained by the boulders. It may seem that I have laboured this description of the beach. There are two reasons why this description is necessary.

The first is that I understand my images better if I know their context. In natural history and landscape photography it seems to me to be important that we understand the dynamics of the objects we are trying to record. In this case the dynamic is water flow and its power to erode.

The second reason is that, from a photographic point of view, it is important to recognise the challenges that these beach conditions present. For example, in many cases the task is to photograph moving water. So, we are back to the familiar debate about shutter speeds. Do we use a fast speed and freeze the flow, or do we let the water run for a bit whilst we keep the shutter open?

This automatically brings us to a consideration of the need for a tripod. If we are using slow shutter speeds then we definitely have to use a tripod. If we want to achieve a precise composition we need to use a tripod anyway.

Furthermore, we cannot make decisions about shutter speed independently of a decision about the focal length at which we want to set the lens. If we want to retain maximum sharpness through a considerable depth of the image then we have to use a small aperture. In order to achieve the correct exposure we then have to use a slow shutter speed, and the decision over how to record the water flow is made for us.

There are two possible alternatives in this situation. The first is to increase the ISO so that we have the freedom to use a faster shutter speed (and achieve a less 'milky' look to the water flow), or acquire a tilt and shift lens so that we do not need to use such a small aperture in order to gain sharpness over the whole image. (Clearly setting a higher ISO is the cheaper option, but there are photographers who would not dream of taking this kind of shot without using their 'tilt and shift'.)

It sometimes helps to include (within one image) both an immovable element, such as a boulder, as well as the flowing water. The presence of the fixed subject enhances the sensation of water movement. There may also be interesting images in the smaller details of the beach. The smaller the area recorded the more the result will approach an abstract pattern. It is important to remain open-minded about the level of detail to record, and to be prepared to be influenced by what lies at our feet. Time of day really matters.

On this particular beach the boulders cast heavy shadows when the sun is coming in at a low angle. This can dominate and spoil an image. Nearer the sea, however, where channel features are very shallow as the water spreads out across the sands, a low-angled morning or late afternoon sun, can help to pick out the details of sand patterns created by the flowing water.

Photographing patterns in the sand may, at first, seem to be a trivial task. In practice it is a significant photographic challenge. The secret of success lies in taking advantage of any light low enough in the sky to create shadows from small features.

When the light is right, choose a good composition, and use a tripod. By using a tripod you will be able to hold on to that composition without the danger of camera shake.

The reward for this care and attention is a collection of fascinating, informative, and decorative images of nature at work (more of that in *Section* 13).

Patterns in the sand as fresh water flows to the sea, Petit Bot Bay, Guernsey.

This is more than just a photograph of flowing water across a beach.

It is true that it tells us about the interplay of sand, rocks and water flow, but it is the arrangement of those components within the image that gives it a pictorial feel. This pictorial sense is enhanced by the colours and reflections. However, there is an error! Have you spotted that the framing as seen through the viewfinder has chopped off a part of the curve cut into the sands by the water flow?

This would probably not have happened if the camera had been mounted on a tripod. The purpose of a tripod is not just to provide stability, it is also serves to keep the camera in one place so that the desired composition is not lost be inadvertent hand-held camera movement.

Graphic (abstract?) patterns, Petit Bot Bay, Guernsey.

Flowing water creating ripples in the sand, Petit Bot Bay, Guernsey.

Note: *both of the above images are examples of where we can photograph the processes that created the patterns as well as the patterns themselves.*

***Pictorial image where water is a major component.
(Isle of Skye, Scotland)***

As Shai Ginott observes in her book *Echoes of Landscape*: "*in landscape photographs one element*" (in the above image it is water) "*assumes centre stage while other (elements) accompany it, playing supporting roles*" (in this case that supporting role is played by the land).

So, are these pictorial images because they are actually closer to what we think of as landscape photographs?

Actually, who cares?

If the viewer likes them, and it satisfies you the photographer, what does it matter which they are?

I think of this image as being pictorial rather than telling me anything about water, and yet for me the water dominates the scene. (Italian Dolomites)

In terms of photographing water-related landscape processes there is more in this image than just a lake and shoreline. We could guess, given the mountain context, that this lake is a legacy from the Ice Age, and that it is dammed behind a glacial terminal moraine.

However, look at the white streaks coming down the mountain side in the distance. These are composed of boulders and rocks that have been shifted partly by gravity (as rock fall or rock slides) and are now caught up in gulleys where torrential flow carries them down the mountain side and spreads them out in alluvial fans. (The less active parts of these alluvial fans are becoming forested.)

*Sunlight through the storm.
(Portree Bay, Isle of Skye, Scotland)*

There is always an element of good fortune in photography. Here the ray of sunshine happened to fall on the boats, as well as near the bottom right-hand corner of the image. In compositional terms this is a good position for a strong element in any picture.

Opportunities like this are almost impossible to plan. You have to be ready to capture them whenever they occur. Similarly with shapes in the water that are momentary events. Seeing and being ready are the two most important elements in such situations.

Ripples made by a passing boat.

Autumn leaves trapped under a flow of water.

13

WATER AT WORK IN NATURE

Water, in several forms, has the power to change the landscape. Several references to this have already been made.

In an earlier discussion we saw how our photographs can pick up the locations where flowing water is at its most turbulent and where its flow is concentrated within a stream. It is in these locations that it has greatest energy and is capable of doing the most erosive work.

Water can reach a river channel through overland flow - i.e. it runs across hillsides in rills and gullies or even as sheets of water in order to reach the main stream or river channels. This flow, if turbulent, can cause soil erosion. We pick that up in our photography when we see murky or muddy water in the stream channel (as in the case of the **River Danube** in the next image).

River Danube near its mouth at the Black Sea.
(Exposure: 1/400th sec at f/11. Lens focal length 200mm., ISO 800)

The River Danube is working as a tool for erosion and as a vehicle for sediment transportation (look at the colour of the water). It has its source nearly 2000 miles away, and throughout its length it carries and deposits sediment according to the state of the river. When in flood much of the sediment load is in suspension.

During low water this sediment is deposited on banks and on its bed. The coarser material is left behind as the river looses its competence to carry gravel. The finer sands silts and clays may be carried all the way towards the river mouth.

Finally the load of sediment is deposited as it approaches the Black Sea. It is spread across the low flat land with every flood, and thereby creates the delta within which the Danube (in this section) now flows.

As water erosion takes place it carves out distinctive landscapes, dramatically seen in parts of south-western USA.

Deep incision as a result of erosion into horizontal strata of bedrock, Canyon Lands, USA

At its extreme this process of deep incision into an old plateau surface is illustrated by the **Grand Canyon of the Colorado**. If you look carefully at the Colorado River in this picture you will see that it is carrying a sediment load of eroded soils (as signified by the brown colour of its waters).

The same process of sediment transport can be seen in the next picture.

This view shows a small section of the Grand Canyon and incorporates the sharp boundary between plateau and gorge, and the Colorado River at the bottom, with its sediment load revealed by the brown colour of the water.

A longer view of the Grand Canyon and the sediment carrying Colorado River.

(The colour shift between the two photographs is because the images were taken at different times of the day on two separate days. It happens. You can either not worry about it or spend lots of time matching them using image processing software - its up to you.)

Photographically it is difficult to know whether to use the camera in portrait or landscape format, so one of each has been included here. There is no right or wrong in matters like this. You have to decide which version works best for you.

In the landscape format version a portion of the uneroded plateau has been included (left-hand side of the image) so that the sharp boundary with the gorge is retained. This edge provides a mental picture of quietness on the plateau and drama in the gorge. It is the contrast in landforms that tells the story of the work of the river (aided by gravity as weathering loosens the soil and rocks on the steep valley sides).

A high view point always provides a good vantage point for photographing deep canyon landscapes, and few places provide a higher viewpoint than that provided by a helicopter. From these images there is a great deal that can be learned about the location of shallows, deposits, concentrations of river flow and the relative densities of sediment load being carried.

In this picture (above) we can see the junction of the Colorado with the Little Colorado (coming in from the left). From the colour of the water it is clear that the smaller of the two rivers is carrying the greater sediment load.

This is not always the case, but it happens to be so at this moment in time because a passing rainstorm fell on the catchment of the Little Colorado but missed that of the Colorado. It is this rainstorm that has initiated further erosion and sediment transport in the Little Colorado.

Again we have been able to use a photograph to tell the story of a process happening over both space and time. This picture shows that soil erosion has happened out of view (to the left) and that the sediment will eventually be deposited to create new landforms further downstream (out of view beyond the edge of the picture).

When this kind of situation happens one photograph can tell us a great deal about the water processes at work in a landscape, and this includes areas well outside the limited scene in the photograph.

Soil erosion is of huge importance to communities that are reliant on the land for agriculture. There are many parts of the world where rural economies depend entirely on produce from the land.

In this context images such as the next one, taken in South Africa, are alarming.

The Augrabies Falls, on the Orange River, South Africa.

The colour of the water shows that the river is carrying a large amount of sediment. This sediment has (very largely) been derived by overland flow and gully flow from the agricultural areas upstream. So, not only is it a dramatic sight it is also a sign that all is not well with the fields and agricultural practices higher in the river catchment area. Soil erosion is at work, and agricultural damage is in process.

Photography now becomes a means of creating an indicator of land management practices.

What has to be determined is the excess of sediment that is derived from man's working of the land over and above that which would have been created by natural processes of erosion.

Throughout the world natural processes of erosion lead to the creation of distinctive landforms. This discussion returns again to North America.

Erosion (coupled with weathering) has also helped to create erosional landscapes such as **Bryce Canyon (above)** and the **Zabriskie Point area (below)**.

The landforms in both of these areas are different from those of the Grand Canyon because they have a different context. Strictly speaking Bryce Canyon is not a canyon because there is no opposite canyon wall. It is an escarpment on the edge of a plateau. This escarpment is retreating quite rapidly because of water and wind erosion of its face.

At Zabriskie Point the rocks are quite different, and more inclined to form gully systems than isolated stacks of rock. If you look closely you will see that the floor of the main gulley is filled with sand and gravel that has been eroded from the hillsides and is waiting for the next heavy rainfall to carry it further down the valley.

These two areas provides golden photographic opportunities as well as telling interesting stories about landscape evolution and the natural sculpturing of the rocks.

The corollary to erosion, of course, is deposition. This river view along the **Redwoods Drive in California** shows banks of deposited sand and gravel on the inside of each meander bend. These deposits have been left behind during the falling stage of a previous flood.

Such deposits also can be left as mid-stream sand banks. When this occurs along a navigable river they become a real hazard to vessels (as here on the **River Danube**).

This theme of water working in the landscape is not restricted to flowing water. When water occurs as ice, and in particular as glaciers, then it too can work in the landscape to create unique landforms. Active glaciers are to be found in many of the higher mountains of the world (e.g. Alps, New Zealand Alps, Alaska, Himalayas, and Rockies). They flow from the higher ice caps down pre-existing valleys eroding as they go.

Fox Glacier, South Island, New Zealand (Exposure: 1/500th sec at f/8. Lens focal length 70mm, ISO 400)

The valley floor beyond the glacier snout is filled with sediment laid down by, and washed out of, the glacier (see photograph below).

Clearly the sediment that is eroded by the glacier has to be deposited somewhere. This may be in the form of moraine or it may be washed out of the glacier as it melts.

The melting ice of the glaciers of the Fritz and Mount Cooke Ranges (South Island, New Zealand) generates outwash channels with the distributaries running through the gravel that has been washed out of the eroding glacier.

Once again we can make our pictures tell a story by selecting the key elements of the glacier at work.

We live in an age when many of the world's glaciers are retreating. 'Retreating' is the wrong word really, for the glaciers are not flowing backwards! Their forward position, however, is progressively changing as the ice melts more quickly than it is replaced by new flows from higher up the valley.

This process, over many years, has left signs of former glaciers and glacial erosion where none now exist. This may be seen in the characteristic U-shaped valleys of such previously glaciated areas.

This photograph (taken in the Swiss Alps) has captured both the U-shaped profile (in the fore-ground) that is the signature of a mountain valley that has been eroded by a glacier, and (in the background) rocks that have been smoothed by glacial erosion as the ice passed across their surface.

In addition, the screes below these rock surfaces consist of lumps of rock that have been prised off the mountainside by the action of frost. Water retained within rock interstices freezes in the winter. As water freezes so it expands in volume. This in turn creates a pressure that separates the lump of rock from its parent. When the ice melts there is nothing left to hold the block in place and it falls down the mountain.

So, when we photograph this type of landscape, we can do so with the purpose of recording those elements that tell us something about the history of the landscape.

We can approach the photography of the coast in a similar manner. It is generally recognised that waves have an erosive power and are capable of transporting sediment. Our photographs of the coast say so much more if we can convey something of that work by the waves. The illustrations used in *Section* 7 showed the power of waves. This power is enough to cause erosion and to carry sediment. We can find sites where these two processes have taken place and photograph them. However, it is sometimes possible to go a step further and record the process itself.

Oregon Coast, USA

In the image (above) erosion of a less-resistant band of rock at the base of the receding cliff has allowed the bay to encroach inland beyond the outcrop in the cliffs at the two ends of the bay.

This erosion operates by under-cutting of the cliffs, causing landslides. The softer material dumped by the landslides on to the beach is then easily removed by the waves which are then left free to erode the base of the cliffs again, causing more landslides, and progressive retreat inland of the cliff line. In this Oregon example, we are recording the results of coastal erosion.

Storm Waves hitting the promenade, Sidmouth, Devon, UK
(Exposure: 1/250th sec at f/16. Lens focal length 95mm. ISO 400)

These waves at Sidmouth were carrying pebbles, striking the promenade wall, and in time are capable of causing damage to such man-made structures. One of the main actions of waves is to carry sediment (sand and gravel) from one location to another. In doing so they can cause erosion in one place and accretion in another.

Beached boat being covered by pebbles, Sidmouth, Devon, UK
(Exposure: 1/80th sec at f/16. Lens focal length 65mm. ISO 100)

Note: *These last two locations are within a mile of each other.*

When such powerful and often erosive work threatens property or life there may be a human reaction such as the building of sea defences.

Sea wall, Southwold, England
(Exposure: 1/320th sec at f/9. Lens focal length 90mm, ISO 200)

The water-related processes described so far have all been physical in nature. It is much harder to record the chemical action of water in the natural environment.

Features resulting from precipitation processes within water at Mono Lake have already been shown (in *Section* 9).

You will have to go underground to photograph another form: stalactites and stalagmites. These are formed as a result of chemical precipitation from ground water within underground caverns.

Stalactites, Blue Mountains, Australia.
(Exposure: 1/13th sec at f/5. Lens focal length 67mm. ISO 400)

Underground photography, with only artificial light available, is very difficult, especially if you are not allowed to use a tripod. This is especially the case when you are in a party that has to keep moving, and there is little time to consider exposures and/or composition.

There are two main problems in the photography of cave features. One is how to set the light balance. My solution is to put it on automatic and sort it out when processing the RAW image.

The other problem is camera shake whilst hand-holding the exposure. The solution is to bump up the ISO as far as you can (without loosing image quality), lower the *f*-value as far as you can without loosing the necessary depth of focus, and make sure the vibration reduction (or its equivalent on your camera) is switched to 'on'.

Above ground some of the most photogenic features formed by chemical processes within ground water are to be found at volcanic sites such as Rotorua (North Island, New Zealand) and Yellowstone National Park (USA).

Mineral deposits, Rotorua, New Zealand.
(Exposure: 1/30th sec at f/8. Lens focal length 105mm. ISO 200)

Unless special permission can be obtained to visit sites, such as the Rotorua Hot Springs, outside normal visiting hours, it will not be possible to use a tripod. Viewing is only permitted from the board walks. These are too narrow and too busy to make the use of tripods a safe option. So, hand-held settings have to be used (i.e. shutter speed fast enough to avoid camera shake, depth of focus sufficient for the location, appropriate ISO value, zoom lens to allow framing without the need to change lenses).

Mineral terraces, Yellowstone.

The same restrictions on the use of a tripod applies to many of the sites within Yellowstone - though I have seen them in use here, and thought in nearly every case "there is an accident waiting to happen".

14

WATER AS A LIVING ENVIRONMENT

We cannot forget that water is important to all forms of life. It may be a threat and a danger in some places, but it is also the basis for life. Not only do we need water in order to survive, it is also the medium on which and within which life exists.

These pelicans need a water environment
(in this case the delta of the River Danube) in order to survive.
(Exposure: 1/320th sec at f/11. Lens focal length 200mm. ISO 800)

Likewise the crocodile and these Hippos in Uganda. Water (in this case the River Nile) is essential to maintaining their life.

Confused Ducks, Whirlow Bridge, Sheffield, UK.
(Exposure: 1/100th sec at f/8. Lens focal length 80mm. ISO 200)

Ducks are an obvious example of birds that require a water environment. However, that does not mean that they are always at peace with each other!

Longshaw, Derbyshire, UK
(Exposure: 1/6th sec at f/16. Lens focal length 97mm. ISO 200)

Note: *A slow shutter speed was chosen in order to capture the movement.*

It would be lovely to say

that I am an under-water photographer, and can bring you pictures of life under the waves. Unfortunately I am not. However, we can all cheat a little. Some quite interesting images can be created within sea-life centres, and it is from these that the next two images come.

When photographing water life in a tank there are several difficulties. These include: low light levels (turn up the ISO setting), reflections in the glass side of the tank (hold the camera up against the glass), condensation within or on the glass sides of the tank (move to another tank).

Blue lobster. Oban Sea Life Centre, Scotland
(Exposure: 1/13th sec. at f/5.6. Lens focal length 70mm. ISO 3200)

Many under-water creatures move very slowly, so shutter speed is seldom an issue. However, sea-life centres are not the places for tripods (unless you can get special out-of-public hours permission). It is neither fair to nor safe for other visitors. You will have to make do with hand-held images. Hence, in this case the high ISO of 3200, and even this high ISO value only allowed a tricky 1/13th sec. exposure in the poor light of the tank Even under these restrictions some interesting images are possible.

Nimmo found!
(Exposure: 1.50th sec at f/4. Lens focal length 105mm. ISO 3200)

At some holiday locations you may have the opportunity of going in a semi-submersible submarine, in which case you will feel almost a part of the life that exists within the water.

Fish within coral reef, West Indies.

15

HUMAN USES OF WATER

Man's adaptation of water for his own use provides us with a rich source of images. Just a few are illustrated here, but you will get the idea very quickly.

Water can be used for either decorative or utilitarian purposes. Many of such uses of water can provide photogenic situations. Let's look at some examples, and how to photograph them.

Take water fountains as an example. Just as with natural landscapes we can record them in their setting or we can concentrate on the way the water moves over and around the fountain. The principles and methods of photography are just the same as those described in earlier sections for flowing and falling water.

Water fountain, Melbourne, Australia Showing the fountain in its context. (Exposure: 1/320th sec at f/5.6. Lens focal length 31mm. ISO 100)

Water jets rising through ornamental paving, Sheffield, England
(Exposure: 1/60th sec at f/11. Lens focal length 40mm. ISO 200)

***Fan arrangement of cascades and fountain, Station Forecourt
Sheffield, England.***
(Exposure: 1/25th sec at f/16. Lens focal length 200mm. ISO 400)

Water feature, Station Forecourt, Sheffield, England
(Exposure: 1/8th sec. at f/16 lens focal length 70mm. ISO 400)

Large water feature, Alnwick Castle, England.
(Exposure: 1/125th sec at f/8. Lens focal length 73mm. ISO 800)

Water-human interaction, Belgrade.
(Exposure: 1/80th sec at f/11. Lens focal length 105mm ISO 400)

Wash day at Lake Garda, Italy.
(I particularly liked the abandoned doll on the beach.)

Emperor Fountain, Chatsworth Park, Derbyshire, England.
(Exposure: 1/100th sec at f/11. Lens focal length 93mm. ISO 200)

Monte Palace Gardens, Madeira.
Sometimes you get a chance to go behind the water feature and can photograph it from an unusual angle.

Lake, Audley End, East Anglia, England.
(Exposure: 1/2000th sec at f/8 Lens focal length 47mm. ISO 1600)
This lake shows the use of water as an ornamental feature within a landscaped garden of a country house (now managed by English Heritage) in England.

Most important of all is the need for water as a human resource. For photographic purposes reservoirs can be treated as though they are lakes.

Howden Reservoir, Derbyshire, England

Donkeys turning the pump at a well, Suez, Egypt.

Though at times the water resource may not be apparent at the surface, we can still photograph the way in which it is being drawn out of the ground, as here in Egypt.

Paddy Fields, Nepal.

Rivers provide water as a resource (as for the paddy fields in which the water has been diverted from a mountain stream). They also provide a means of transport.

Early morning barges on the River Rhine upstream of Bonn, Germany.
Exposure: 1/100th sec at f/11. Camera focal length 105mm. ISO 100)

The site of Koblenz at the confluence of the Rivers Rhine and Moselle – a major calling point for barges.

(The name 'Koblenz' is derived from the same stem as the word 'confluence'.)
(Exposure: 1/60th sec at f/11. Lens focal length 70mm. ISO 100)

The major rivers of the world provide important routes for the transport of goods, the siting of towns, and the locations for industry, especially those that either require the water or the transport facility provided by the river. They also provide many photographic opportunities.

Storage facility and barge traffic along the River Rhine.
(Exposure: 1/50th sec at f/11. Lens focal length 85mm. ISO 100)

Inland waterways are to be found within some major towns and cities. The obvious example is Venice, but other places also have extensive urban waterways, including Copenhagen and Strasbourg.

Urban waterway, Copenhagen.

(Exposure: 1/80th sec at f/11 Lens focal length 28mm. ISO 400)

Strasbourg, France.
(Exposure: 1/200th sec at f/7.1 Lens focal length 100mm. ISO 200)

Canals were often designed to operate between major rivers. Many were created to provide transport for goods to and from industrial areas. In some cases (e.g. The Netherlands) they also serve (and sometimes principally serve) as a means of controlling land drainage. Today many of these canals and rivers are also used for leisure purposes.

Canal Long Boat passing out of the River Soar and into the River Trent (England): the use of water for transport.
(Exposure: 1/40th sec at f/18. Lens focal length 55mm. ISO 1000)

Note: *the cooling towers for the power station in the background. This demonstrates another use for water.*

There are many other forms of leisure uses of water, each of which can be a photo opportunity. Obvious examples include coastal resorts.

Marigot Bay, St Lucia
(Exposure: 1/350th sec at f/10 Lens focal length 23mm ISO 160)

Gulf de Porto Vecchio, Corsica

In addition to recording such places we can create images out of the leisure activities.

Children enjoying the sea, Madeira.

Beach scene, Cornwall, England
(Exposure: 1/640th sec at f/9 Lens focal length 100mm. ISO 400)

On the other hand if it is a bad day:

Empty deck chairs, Beer, England
(Exposure: 1/250th sec at f/16 Lens focal length 85mm ISO 400)

Some leisure activities on water are more vigorous than others:

Jet Skier on the River Rhine
(Exposure: 1/250th sec at f/8 Lens focal length 105 mm ISO 100)

Boating off St Lucia

Ladybower Reservoir, Derbyshire, England.
(Exposure: 1/250th sec at f/9 Lens focal length 55mm, ISO 400)

   A common pastime, in fact the sport with the greatest number of individual participants in England, is fishing.

Sea fisherman, Sidmouth, England
(Exposure: 1/50th sec at f/16 Lens focal length 55mm ISO 100)

Boats at anchor can also provide photographic opportunities:

Fishing Boats, Camara de Lobo, Madeira

Moored boats at low tide, Staithes, Yorkshire, England.

Not all leisure activities are on water as a liquid. Some are quite happy to use water in its more solid form. Snow and ice sports take on many forms, including just the simple fun of being a child again.

Children sledging, Dore, Sheffield, UK
(Exposure: 1/125th sec at f/16 Lens focal length 100mm ISO 400)

Human uses of water, where water becomes a commercial asset, includes power generation. Hydro-electric power stations provide power from water. The photogenic bit, (as far as water is concerned) is the often to be found just below the dam.

Owen Falls Dam, Jinja, Uganda.

Another form of energy derived from water is to be found in geothermal areas (e.g. Iceland and (in the example below) New Zealand) where underground sources of steam are used for heating or for power.

North Island, New Zealand
(Exposure: 1/800th sec at f/4.5. Lens focal length 200mm. ISO 200)

As the Section on Water at Work (*Section* 13) showed, water can be a hazard as well as a resource. Along the coast the reaction has been to create see defences as sea walls or as groynes.

Sea defences, Southwold, Suffolk, England
(Exposure: 1/160th sec at f/11 Lens focal length 70mm ISO 200)

Concentration of groynes east of Beachy Head, Eastbourne, UK
(Exposure: 1/125th sec at f/16 Lens focal length 110mm ISO 400)

I am sure that you can think of many other interactions between people and water. Good, for these give you other photographic opportunities. Here are two final examples to help you to think of some examples of your own.

The English castles were normally surrounded by a moat, for defensive purposes. The next photograph shows the use of water as a pictorial element in the landscape. This 'moat' use is a long way from a means of defending the castle.

The defensive moat at Hever is up against the castle wall (beyond the dark hedge). This one is just for landscaping and aesthetic purposes.

Moat, Hever Castle, England
(Exposure: 1/30th sec at f/16 Lens focal length 35mm ISO 400)

Finally, an image that includes a combination of the some of the themes discussed in this book.

Here we have clouds, rain in the distance, and a water surface (***Port Philip Bay, Melbourne, Australia***). We can also see how the water is being used by a lone fisherman. (*Exposure: 1/15th sec at f/6.3 Lens focal length 135mm ISO 400*)

This sequence of images illustrates the many good opportunities that exist for photographing water. The theme throughout has been: - take advantage of photogenic light - use the shutter speed knowing its effect on the final image, especially if water is either flowing or falling. - use a tripod whenever it is appropriate to do so, and use it for composition as well as for controlling camera shake.

The following (concluding) Section is provided for those who would like a little more discussion of two techniques mentioned in the above text. These are:

– how to use a Neutral Density filter to help to extend the exposure time when photographing moving water and
– how to apply HDR (High Dynamic Range) techniques.

16

TECHNICAL NOTES

Extending exposure time

There is one technical point that needs to be tidied up. Some play has been made on the representation of water using long exposures. Thereby creating an image of flowing water which renders it in a way that we do not see with the human eye. We see in 'real time' the long shutter time records it in 'extended time'.

The customary practice when recording in extended time is to set the camera on a tripod and choose a small aperture (say f/22) in combination with a low ISO value. This automatically requires the use of a slow shutter speed especially if the ISO is turned down to 100.

The exposure time can be further extended by adding filters to the camera lens. These are usually in the form of neutral density (ND) filters of various strengths. These ND filters serve to reduce the amount of light reaching the camera sensor, hence a longer exposure is required. The most extreme form of these in general use is the ND400. This filter is so dark that it is impossible to mount it on the front of the lens and still expect to be able to manually focus correctly. In practice focussing has to be carried out (in manual mode) before the ND400 is screwed in front of the lens.

Also difficult is the estimation of the correct exposure time. You will need to put the camera on Manual and then use the Bulb Setting. It is this long exposure that causes moving water to be recorded as a smooth surface (sometimes with a 'milky' look).

Use an aperture of f/11 or f/16 (with the ISO set at 100) and place the camera on a firm tripod. Establish your composition, take a test image, and establish what your normal camera shutter speed would be in order to provide you with a good exposure.

Now comes the tricky part. Place the ND400 filter over the lens (be careful not to change the focus). Use the Bulb Setting and expose the image for a time which is about 1000 times longer than it would have been in the absence of the filter. (You will need to judge the length of exposure using a watch.)

This means that you will use the following settings:

If the 'normal' setting is:	then the new setting is:
1/500th sec	2 sec
1/250th sec	4 secs
1/125th sec	8 secs
1/60th sec	16 secs
1/30th sec	32 secs
1/15th sec	64 secs

Once you have taken the exposure study it, together with its histogram, and decide whether the exposure time was correct. If not, adjust this time and repeat until you get the exposure right.

Is it all worth it?

Only you can be the judge. So, to what can you look forward? I will give you a few examples, then you can decide if the resulting images suit your taste.

*A close-up of flow over rocks within a stream showing the effect of using a ND400 filter.
(Burbage Brook, Derbyshire, UK.)*
(Exposure: 20 secs at f/10 Lens focal length 50mm ISO 800)

A wider view of a similar site:

***A section of Burbage Brook (Padley Gorge), Derbyshire,
England - using a ND400 filter.***
(Exposure: 30secs at f/16 Lens focal length 50mm, ISO 800)

The same technique can also be applied to the sea:

This view of the *Isle of Skye, Scotland*, uses a ND400 filter to create a smooth surface over the sea. (*Exposure: 15 secs at f/22 Lens focal length 105mm ISO 100*)

It is highly likely that your early attempts to use a really dark filter like this will lead to some disappointment. It can take several attempts before you achieve a decent image. It is great fun, but highly frustrating.

The truth is, we are all learning, and even with practice we may not become perfect. However, we will become better!

HDR photography

There have been several references in this book to HDR - High Dynamic Range - photography. This techniques is used when the range of exposure values within an image is so great that it exceeds that which can be recorded by the camera sensor. In other words, highlights get blown out and shadows lack detail.

In order to over-come this problem, images are recorded at several exposures, making sure that all of the detail within the highlights are recorded in at least one image, and all of the shadow details are captured in another. Several other exposures can/may be taken in-between these two extremes. It is important to do this whilst the camera is mounted on a tripod. This ensures that each successive image is of exactly the same subject.

Special software is then required that uses the information from each of this succession of images and combines them so that maximum detail is retained (within a final combined image) for both the highlights and the shadows, as well as everything in-between.

The following image is the result of combining three separately exposed images and processing them using HDR software. Using the HDR technique has allowed detail to be shown within the surfaces of the darker rocks as well as within the brighter areas of water flow. This is a very useful tool in areas of heavy shadows in an otherwise bright scene.

HDR applied to fast flowing stream, Padley Gorge, Derbyshire, England.

Note: This image uses a combination of HDR and ND400 filter techniques.

WATER - A FINAL COMMENT.

I hope you have enjoyed this insight into the photography of water.

In this book you have seen some of my methods and the results. Now it is up to you. Copy my techniques if you want to do so, but try and develop methods and ideas of your own.

The subject is without limits. Your imagination should be the same. If this book has stimulated some new ideas for your photography, then it has achieved something.

Just don't all get the same idea at the same time, or it could get crowded out there!

QUICK SUMMARY

For many pictures the techniques to be used require the same normal care and attention that would be applied to any serious image:
1. use a tripod whenever possible
2. if hand-held make sure the shutter speed is sufficient to overcome camera shake
3. the aperture setting must be appropriate for the depth of field (focus) required
4. the ISO setting should be kept low enough to avoid pixelation within the image.
5. Special techniques for recording water flow tend to revolve around the shutter speed setting on the camera i.e. the slower the speed the more "milky" the appearance of the water. In recent years this has been extended by the use of a ND400 filter.
6. Benefits can be derived from the use of a polarising filter, especially to reduce reflections and highlights from the surface of the water.
7. Other special steps could include the use of HDR (High Dynamic Range) techniques in order to over-come situations with bright highlights and deep shadows. (Though a great deal can be done to counteract extremes of brightness by recording a RAW file and within the RAW processing reduce the exposure value and increase the brightness value. This has the effect of 'saving' many areas that otherwise would have been detrimental to the image.)

Summary of points raised in the text:

1. Water can exist in solid (ice, snow), liquid, or vapour forms (clouds, steam).

2. Water can occur in a variety of states (e.g. calm, turbulent, flowing, falling, breaking, splashing).

3. The form in which water occurs can determine the way we approach the photography of water.

4. As photographers our main consideration is whether the water is moving or is static.

5. Static water can be photographed using the same techniques as for landscape photography.

6. Moving water requires special care over shutter speed. You need to decide if you like your water to look milky or as if it is frozen, or somewhere in between. Your choice of shutter speed and whether or not you want to use a ND400 (or similar) filter will determine this.

7. As in all photography, keep it simple. Don't over clutter your picture.

8. Choose your angle of view with care.

9. Home in on detail when that detail makes a worthwhile image in its own right.

10. Practice by starting with a general view and gradually move in until you are getting down to the real detail.

11. A polarising filter is invaluable in removing unwanted glare from the surface of water.

12. Be careful not to over-do the polarising effect as this can render the colour of the sky an artificially deep blue.

13. A slow shutter speed will remove any gentle ripples that occur on the surface of a lake and show it as a smooth water surface with a soft tone in the image.

14. If the water is being used for a particular activity (e.g. kayaking over rapids) it may be worth using your picture to tell a story by including people engaging with the water.

15. Many clouds have a form and texture that are well worth recording in their own right.

16. Under expose a white cloud in order to retain any detail within its form.

17. Clouds can be the subjects of almost abstract images.

18. Photographically clouds can be used to advantage when they

are intermingling with peaks and mountains.

19. When it comes to the photography of individual raindrops it is necessary to use the techniques associated with macro-photography.

20. Falling rain provides a photo opportunity. You can show splashes in a puddle, streaks of rain across a view (use a slow shutter speed), rain falling on a window or a car windscreen, and so on.

21. Although flowing water can look good in a general view, the best pictures may well be in the detail of water flow.

22. In a flowing river look for: flow patterns, interplay with exposed rocks, and any water splash. (These are usually associated with fast flowing streams in hilly and mountainous areas.)

23. Light falling on the water surface can add a significant dimension to your image.

24. The turbulent upper crest of a waterfall, or its base, are prime locations for water images.

25. A telephoto lens may be required in order to pick out small details of river flow.

26. When bright light is reflected from the surface of the water exposure calculation becomes tricky. You cannot rely on a general reading. Take a spot reading on a mid-tone area and use that. Bracket your images by taking two more shots, one on each side of the spot reading (using +0.3EV and -0.3EV).

27. The best lighting conditions, with far fewer exposure difficulties, occur with a bright but over-cast sky rather than bright sunshine.

28. Remember to use your camera in portrait format if your subject warrants it.

29. There are times when you need to show the environmental setting as it provides a valuable context for your water picture.

30. Catching a breaking wave at just the right moment is very hard. Set your camera on "continuous shooting" and choose the best image later.

31. Still and calm water implies "peace" and our photographs need to pick that up. One way to do that is to have a sharp reflection in the water.

32. Saline lakes form a special environment (and may be particularly good for wildlife photography).

33. Volcanic geysers are best taken against a sky of contrasting

colour.

34. When photographing snow or ice bracket your exposures to make sure at least one image is correctly exposed.

35. Let images show water at work (e.g. in sediment transport and erosion).

36. Photographing the uses made of water by mankind can be an interesting photographic project in itself. But the principles of photography are the same as they are for water in its natural forms.

37 Some images of water (in whatever form) can be rendered as abstract images, even if at their heart they are pictorial.

38 Use a tripod in order to have tight control over image composition as well as for keeping the camera steady on long exposures

ABOUT THE AUTHOR

I got hooked on photography back in 1956 when, as a student, I acquired my first (very cheap) basic 35mm film camera. Since then I have progressed through a range of cameras whilst also travelling the world as a professional geographer. Although my main interest has always been in landscape photography, of necessity I have also gained wide experience in many other aspects of photography.

Over the past 20 years I have wanted to pass on some of that experience and have been fortunate enough to have articles accepted by *Amateur Photographer*, *Outdoor Photography*, as well as *Practical Photography*, *Freelance Photography Made Easy*, and *f2 Freelance Photographer*.

With the growing popularity of Kindle I decided to bring all of my experience and enthusiasm for photography into one substantial series of books. However, I wanted to write for my many friends who are keen to be better photographers but do not aspire to become Professionals.

So, this Series is for those of you who want to learn more and learn quickly about the art and science of photography without "drowning in the deep end".

The only good photographer is the one who enjoys it, has an 'eye' for a photogenic situation, and always wants to improve their technique.

Out of that enjoyment and well-applied techniques will come creativity and great pictures. One Last Thing Amazon gives you the opportunity to rate this book. If you believe that it is worth sharing please take a few seconds to tell others. If they too become better photographers as a result they will always be grateful to you. As will I. Happy and creative photography!

John C Doornkamp

(www.doornkamp.co.uk)